HOW TO BALANCE LIVE SOUND PRODUCTION FOR BEGINNERS

ESSENTIAL GUIDE AND TECHNIQUES , STRATEGIES FOR ACHIEVING OPTIMAL LEVELS IN A LIVE MIX

Copyright@2024

Speck Asencio

TABLE OF CONTENT

- CHAPTER 1: BASICS OF SOUND 9
  - WHAT IS SOUND? 9
- CHAPTER 2: EQUIPMENT OVERVIEW 33
  - MICROPHONES 33
- CHAPTER 3: SETTING UP A LIVE SOUND SYSTEM 80
  - STAGE LAYOUT 80
- CHAPTER 4: SOUND CHECK AND LINE CHECK 102
  - IMPORTANCE OF SOUND CHECK 102
- CHAPTER 5: BALANCING THE MIX 112
  - UNDERSTANDING THE MIX 112

INTRODUCTION

Purpose of the Book

Live sound production is the art and technological know how of dealing with and enhancing the audio enjoy of live activities, from concerts and theater productions to corporate activities and conferences. This ebook is designed to provide beginners with a comprehensive guide to information and imposing the ideas of live sound production. By the stop of this ebook, you may have a strong basis inside the abilities and expertise required to create a balanced and expert live sound mix.

This e book is aimed toward aspiring sound engineers, musicians, occasion organizers, and everyone interested in studying the basics of stay sound production. Whether you're a entire beginner or a person with a bit of enjoy looking to refine your capabilities, this guide will help you navigate the complexities of stay sound. No earlier know how is assumed, and the content is provided in an clean to apprehend way with plenty of sensible examples.

HOW TO USE THIS BOOK

To get the maximum out of this ebook, we advocate the following technique:

Start from the start: The chapters are established to construct upon every different, so starting from the start will ensure you hold close the fundamental ideas before shifting directly to greater advanced topics.

Practice frequently: Live sound manufacturing is a finger son area. Whenever feasible, follow what you examine by means of training with actual gadget and stay scenarios.

Use the sources: Throughout the e book, you may locate sensible suggestions, diagrams, checklists, and actual global examples. Take advantage of these assets to deepen your knowledge.

Join a community: Engaging with other beginners and experts in the area can provide additional insights and support. Consider becoming a member of online forums, attending workshops, or volunteering at stay occasions.

UNDERSTANDING THE IMPORTANCE OF LIVE SOUND PRODUCTION

Live sound production is a vital factor of any stay occasion, from small club performances to large scale concerts and festivals. It includes the technical and innovative strategies required to deliver top notch sound to the target audience, making sure that every note, word, and impact is heard definitely and with the favored impact.

1. Enhances the Audience Experience

A properly balanced stay sound mix can appreciably enhance the target audience's universal enjoy. Clear and dynamic sound helps convey the emotion and strength of the overall performance, making it greater engaging and noteworthy. Conversely, negative sound best can detract from even the maximum talented performers, leading to a disappointing enjoy for the audience.

2. Supports the Performers

Good live sound production affords performers with the self assurance to deliver their best overall performance. With proper monitoring and a balanced mix, performers can hear themselves and their fellow musicians certainly, letting them stay in sync and explicit their artistry efficaciously. This support is crucial for keeping the float and great of the performance.

3. Ensures Technical Quality

Live sound production includes handling diverse technical elements, consisting of microphone placement, signal routing, and sound device setup. Attention to those info ensures that the sound satisfactory remains steady and loose from troubles like remarks, distortion, and sign dropouts. Technical talent allows keep the integrity of the audio sign from the supply to the target audience.

4. Adapts to Different Venues

Each live occasion venue gives particular acoustic demanding situations, from small clubs with reflective surfaces to open air gala's with variable weather situations. Live sound production calls for the capability to evolve to those unique environments, optimizing the sound gadget to fit the unique characteristics of the venue. This adaptability is prime to delivering a extraordinary audio revel in irrespective of the place.

5. Balances Artistic and Technical Elements
Live sound production is both an art and a technological know how. It entails creative choices, inclusive of the way to balance specific gadgets and vocals within the mix, as well as technical obligations like setting up device and troubleshooting troubles. This stability among inventive and technical elements is what makes stay sound manufacturing a completely unique and rewarding field.

6. Creates a Professional Atmosphere
Professional sound manufacturing contributes to the overall professionalism of an occasion. High exceptional sound can elevate the perceived price of the occasion, making it more enjoyable for the audience and extra prestigious for the performers. This professionalism also can appeal to higher expertise and large audiences within the destiny.

7. Enhances Safety and Communication

In addition to the creative and technical elements, live sound manufacturing performs a function in making sure protection and verbal exchange for the duration of an event. Clear bulletins and emergency commands are crucial in dealing with big crowds and making sure absolutely everyone's protection. A reliable sound device allows facilitate smooth communique between event staff and the target audience.

CHAPTER 1: BASICS OF SOUND

WHAT IS SOUND?

UNDERSTANDING SOUND WAVES

Sound is a form of power that travels thru the air (or different mediums) as vibrations. These vibrations create stress waves that our ears interpret as sound. To efficaciously stability stay sound production, it is crucial to understand the fundamental residences of sound waves.

1. What is a Sound Wave?

A sound wave is a longitudinal wave that travels thru a medium (consisting of air, water, or solids) because of the vibration of debris in that medium. When an item vibrates, it reasons the encompassing debris to vibrate as well, developing a wave of alternating high and coffee pressure areas that circulate outward from the source.

2. Frequency

Frequency refers to the number of wave cycles that arise in one second, measured in Hertz (Hz). It determines the pitch of the sound:

Low Frequency: Sounds with low frequency (e.G., bass) are heard as low pitched sounds.

High Frequency: Sounds with high frequency (e.G., treble) are heard as excessive pitched sounds.

Humans generally listen frequencies ranging from 20 Hz to 20,000 Hz.

3. Amplitude

Amplitude is the peak of the sound wave and determines the loudness of the sound. It is measured in decibels (dB):

Higher Amplitude: Louder sounds have better amplitudes.

Lower Amplitude: Softer sounds have lower amplitudes.

4. Wavelength

Wavelength is the space among successive crests (or troughs) of a wave. It is inversely related to frequency:

Long Wavelength: Low frequency sounds have long wavelengths.

Short Wavelength: High frequency sounds have short wavelengths.

5. Waveform

The waveform represents the shape of the sound wave and impacts the timbre or color of the sound:

Sine Wave: Pure tone and not using a harmonics, easy and clear.

Square Wave: Contains ordinary harmonics, sounds more buzzy.

Triangle Wave: Contains odd harmonics with a smoother, greater mellow sound.

Saw tooth Wave: Contains each even and ordinary harmonics, has a shiny and edgy sound.

6. Propagation of Sound

Sound waves propagate via special mediums at varying speeds:

Air: Approximately 343 meters in keeping with 2nd (m/s) at room temperature.

Water: Faster than in air, around 1,480 m/s.

Solids: Generally quicker than in drinks and gases, dependent on the density and elasticity of the material.

7. Reflection, Absorption, and Diffraction

Reflection: Sound waves jump off surfaces. Hard surfaces (e.G., walls, flooring) replicate sound, which could create echoes.

Absorption: Soft substances (e.G., carpets, curtains) absorb sound waves, decreasing mirrored image and assisting control room acoustics.

Diffraction: Sound waves bend round boundaries and unfold out after passing thru openings, allowing sound to fill a space.

FREQUENCY, AMPLITUDE, AND WAVEFORM

Understanding the key homes of sound waves—frequency, amplitude, and waveform—is fundamental to getting to know live sound production. These houses influence how we perceive sound and how we will control it to acquire the desired audio satisfactory in a live placing.

1. Frequency

Definition:

Frequency is the variety of wave cycles that pass a set point in one second. It is measured in Hertz (Hz).

Impact on Sound:

- Pitch: Frequency determines the pitch of the sound. Higher frequencies produce better pitched sounds, whilst lower frequencies produce lower pitched sounds.

- Low Frequencies: Typically among 20 Hz and 250 Hz. Examples encompass bass guitars and kick drums.
- Mid Frequencies: Typically between 250 Hz and 4,000 Hz. Examples encompass human speech and maximum musical gadgets.

High Frequencies: Typically between four,000 Hz and 20,000 Hz. Examples consist of cymbals and a few string devices.

Practical Application:

Equalization (EQ): Adjusting the frequency stability of the audio to decorate clarity, remove muddiness, or upload brightness.

Crossover Networks: Dividing audio alerts into extraordinary frequency bands to send them to appropriate speakers (e.G., subwoofer for low frequencies, tweeters for high frequencies).

2. Amplitude

Definition:

Amplitude is the peak of the sound wave, which corresponds to the amount of electricity it includes. It is measured in decibels (dB).

Impact on Sound:

- Loudness: Amplitude determines the loudness or quantity of the sound. Higher amplitude effects in louder sounds, even as decrease amplitude consequences in softer sounds.
- High Amplitude: Loud sounds, consisting of a rock concert or a jet engine.
- Low Amplitude: Soft sounds, which include a whisper or heritage track.

Practical Application:

- Volume Control: Adjusting the general amplitude to make certain the sound is at a appropriate degree for the target audience.

- Dynamic Range Compression: Reducing the dynamic range to make quieter sounds louder and louder sounds quieter, ensuing in a extra regular extent level.

3. Waveform

Definition:

The waveform is the form of the sound wave, which impacts the timbre or man or woman of the sound. Different waveform create distinctive sound qualities.

Common Wave forms:

Sine Wave:

　Description: A pure tone with out a harmonics.

　Sound: Smooth and clean.

　Usage: Often utilized in test tones and synthesizers.

Square Wave:

Description: Contains a essential frequency and extraordinary harmonics.

Sound: Buzzy and harsh.

Usage: Common in electronic song and positive synthesizer sounds.

Triangle Wave:

Description: Similar to a sine wave however with extra bizarre harmonics.

Sound: More mellow than a rectangular wave, however nonetheless truly buzzy.

Usage: Often utilized in digital music and for creating synthetic bass sounds.

Saw tooth Wave:

Description: Contains both even and extraordinary harmonics.

Sound: Bright and edgy.

Usage: Widely used in electronic track for its rich harmonic content.

Practical Application:

Sound Design: Choosing and shaping waveform to create preferred sounds in song and results.

Harmonic Content: Understanding how special waveform affect the harmonic content material of a valid, which can be vital for mixing and EQ decisions.

HOW SOUND TRAVELS

Sound is a type of electricity that propagates via a medium as a wave. To apprehend how sound travels, it's essential to explore the mechanisms of sound wave propagation,

1. Basic Principles of Sound Propagation

Sound Waves:

Sound waves are mechanical waves created by means of vibrating items. These vibrations motive close by debris in a medium (such as air, water, or solids) to vibrate, developing areas of compression and rarefaction.

Compression: A area where particles are close collectively.

Rarefaction: A location where particles are spread apart.

Longitudinal Waves:

Sound waves are longitudinal waves, meaning the particles of the medium vibrate parallel to the direction of the wave's propagation.

Speed of Sound:

The speed of sound varies depending at the medium. In air at room temperature, sound travels at approximately 343 meters in line with 2d (m/s). In water, it travels quicker (approximately 1,480 m/s), and in solids, it travels even faster (varies based on the cloth).

2. Propagation Through Different Media

Air:

Most not unusual medium for sound propagation in live sound production.

Speed of sound is tormented by temperature, humidity, and air pressure. Warmer temperatures and better humidity can boom the speed of sound.

Water:

Sound travels faster and farther in water than in air due to the higher density of water molecules.

Common in underwater acoustics and sonar packages.

Solids:

Sound travels fastest in solids because debris are extra tightly packed.

Used in programs like seismic studies and vibrations in structures.

3. Reflection, Absorption, and Diffraction

Reflection:

When sound waves stumble upon a surface, they could get better, growing an echo or reverberation.

Hard, easy surfaces (e.G., walls, flooring) generally tend to reflect sound more efficaciously.

Absorption:

When sound waves come across a porous or soft fabric, part of the sound strength is absorbed and transformed into heat.

Materials like carpets, curtains, and acoustic panels are used to lessen reflections and manipulate room acoustics.

Diffraction:

Sound waves can bend around obstacles and spread out after passing through openings.

Allows sound to tour via doorways or around corners, ensuring it is able to fill a space.

4. Interference

Constructive Interference:

Occurs whilst sound waves meet in phase, amplifying the sound. Peaks and troughs align, ensuing in a louder sound.

Destructive Interference:

Occurs when sound waves meet out of segment, canceling each different out. Peaks align with troughs, decreasing or getting rid of the sound.

Practical Application:

Understanding interference is critical for speaker placement and microphone setup to keep away from feedback and reap a balanced mix.

5. Practical Implications for Live Sound Venue Acoustics:

The length, shape, and substances of a venue significantly impact sound propagation.

Acoustic remedy (e.G., diffusers, absorbers) can decorate sound best through controlling reflections and reverberation.

Speaker Placement:

Strategic speaker placement ensures even sound coverage and minimizes undesirable reflections and interference.

Line arrays and allotted speaker structures help achieve steady sound ranges during the venue.

Microphone Techniques:

Proper microphone placement and choice reduce remarks and capture sound correctly.

Directional microphones (e.G., cardioid, supercardioid) are used to cognizance on sound from particular assets at the same time as minimizing heritage noise.

PROPAGATION OF SOUND

Sound propagation is the technique with the aid of which sound waves travel thru a medium. This entails the movement of power from the sound source to the surrounding environment. Understanding sound propagation is crucial for powerful stay sound production as it influences how sound behaves in exclusive environments and the way we can manage it for premier audio high quality.

1. Basics of Sound Propagation

Sound Waves:

Sound waves are mechanical vibrations that travel through a medium (consisting of air, water, or solids).

These waves are longitudinal, that means the particle displacement is parallel to the direction of wave propagation.

Compression and Rarefaction:

Sound waves encompass alternating regions of compression (wherein particles are near collectively) and rarefaction (in which debris are spread aside).

This creates stress variations that propagate thru the medium.

Speed of Sound:

The velocity of sound varies depending at the medium and its residences, consisting of temperature, density, and elasticity.

Air: Approximately 343 meters in step with 2d (m/s) at room temperature.

Water: Approximately 1,480 m/s.

Solids: Faster than in drinks and gases; varies with material homes.

2. Factors Affecting Sound Propagation Medium:

The sort of medium drastically affects sound pace and conduct.

Air: Common medium for ordinary sounds and live sound manufacturing.

Water: Sound travels quicker and further in water because of its better density.

Solids: Sound travels quickest in solids because debris are more closely packed.

Temperature:

Higher temperatures growth the velocity of sound because debris move quicker and transmit vibrations extra speedy.

In stay sound settings, outdoor activities may revel in variations in sound pace because of changing temperatures.

Humidity:

Increased humidity commonly will increase the velocity of sound because wet air is much less dense than dry air.

Air Pressure:

Higher air pressure can slightly growth the rate of sound, but its effect is much less sizable than temperature and humidity.

3. Reflection, Absorption, and Diffraction

Reflection:

Sound waves can jump off surfaces, inflicting reflections. Hard, easy surfaces reflect sound more effectively.

Reflected sound can create echoes and reverberation, influencing how sound is perceived in a area.

Echo: A awesome, behind schedule mirrored image of sound.

Reverberation: A collection of reflections that blend together, growing a persistence of sound after the unique sound stops.

Absorption:

When sound waves come upon soft or porous materials, part of the sound strength is absorbed and transformed into heat.

Materials like carpets, curtains, and acoustic panels help take in sound, lowering reflections and controlling room acoustics.

Diffraction:

Sound waves can bend around limitations and unfold out after passing via small openings.

Diffraction lets in sound to fill a area and guarantees that it is able to be heard even if there are obstacles in the manner.

4. Interference

Constructive Interference:

Occurs while sound waves meet in segment (peaks align with peaks, and troughs with troughs), resulting in a louder sound.

Can be used to enhance sure frequencies or acquire specific sound consequences.

Reflection, Absorption, and Diffraction

Understanding how sound interacts with distinct surfaces and boundaries is crucial for effective live sound manufacturing. These interactions are usually categorized into reflection, absorption, and diffraction. Each of these phenomena affects how sound travels and the way it's miles perceived by way of the target market.

1. Reflection

Definition:

Reflection takes place when sound waves soar off a surface and alternate direction. The attitude of incidence (the angle at which the sound wave hits the floor) equals the angle of reflection (the perspective at which it leaves the floor).

Key Points:

Hard Surfaces: Materials consisting of concrete, glass, and metallic reflect sound successfully, often causing echoes and reverberation.

Soft Surfaces: Soft materials like curtains and carpets absorb more sound and mirror much less, lowering echoes.

Echo: A wonderful, not on time reflection of sound. Occurs in massive areas or while sound waves reflect off distant surfaces.

Reverberation: A series of overlapping reflections that blend collectively, growing a staying power of sound after the original sound stops. Occurs in enclosed spaces and is motivated by way of the room's length, form, and floor materials.

Practical Application:

Managing Reflection: Using acoustic treatments like diffusers and absorbers to govern reflections and enhance sound readability.

Speaker Placement: Positioning speakers to reduce undesirable reflections and gain a balanced sound distribution.

2. Absorption

Definition:

Absorption occurs whilst sound waves are absorbed via a material, converting the sound strength into heat and reducing the quantity of sound that is meditated returned into the gap.

Key Points:

Absorptive Materials: Materials including foam panels, carpets, heavy curtains, and acoustic tiles are designed to soak up sound. Their effectiveness is measured by means of the Noise Reduction Coefficient (NRC), which tiers from zero (no absorption) to at least one (entire absorption).

Frequency Dependence: Different substances take in special frequencies. For instance, thick, dense substances are better at absorbing low frequencies, even as softer substances are powerful at absorbing high frequencies.

Practical Application:

Acoustic Treatment: Installing absorptive substances in strategic places to control room acoustics, lessen reverberation, and beautify sound clarity.

Bass Traps: Specialized absorbers designed to manage low frequency sound electricity in corners and alongside walls.

3. Diffraction

Definition:

Diffraction takes place when sound waves come across an obstacle or pass via a gap and bend round it. This allows sound to unfold out and fill a space, despite the fact that there are obstacles.

Key Points:

Obstacles and Openings: Sound can bend round boundaries like pillars and walls or spread out after passing through doors and windows.

Frequency Dependence: Lower frequency sounds (with longer wavelengths) diffract more successfully than higher frequency sounds (with shorter wavelengths).

CHAPTER 2: EQUIPMENT OVERVIEW

MICROPHONES

TYPES OF MICROPHONES: DYNAMIC, CONDENSER, AND RIBBON

Microphones are critical equipment in stay sound manufacturing, and selecting the right kind can appreciably impact sound fine and overall performance. The three foremost varieties of microphones are dynamic, condenser, and ribbon microphones. Each type has specific traits and is desirable for distinctive applications.

1. Dynamic Microphones

Construction:

Dynamic microphones use a diaphragm connected to a coil of twine, positioned in the magnetic discipline of a permanent magnet. When sound waves hit the diaphragm, it actions the coil, producing an electrical signal.

Characteristics:

Durability: Dynamic microphones are sturdy and can handle high sound stress stages (SPL) without distortion. They are less sensitive to moisture and bodily harm, making them perfect for stay performances and rugged environments.

Lower Sensitivity: They have decrease sensitivity compared to condenser microphones, because of this they are less probably to choose up ambient noise and are better at handling loud sound resources.

Applications:

Vocals: Popular selections for live vocal performances (e.G., Shure SM58) due to their durability and capability to address excessive SPL.

Instruments: Suitable for miking loud units like drums (e.G., snare, kick drum) and guitar amplifiers.

Advantages:

Rugged and durable

Affordable

Handles excessive SPL properly

Disadvantages:

Less sensitive

Limited frequency reaction compared to condensers

2. Condenser Microphones

Construction:

Condenser microphones have a diaphragm located very near a back plate, growing a capacitor. When sound waves hit the diaphragm, it adjustments the gap between the diaphragm and the back plate, generating a trade in capacitance that is transformed into an electrical sign. They require an external strength supply, typically phantom power (48V).

Characteristics:

High Sensitivity: Condenser microphones are more sensitive than dynamic microphones, taking pictures greater element and higher frequencies. This makes them best for studio recordings and shooting diffused nuances.

Extended Frequency Response: They offer a much broader and greater accurate frequency response, making them suitable for loads of sound sources.

Applications:

Vocals: Widely utilized in studio settings for recording vocals because of their readability and element (e.G., Neumann U87).

Instruments: Ideal for shooting acoustic gadgets, pianos, and overheads for drums.

Pickup Patterns

Pickup patterns, also called polar patterns, describe the sensitivity of a microphone to sound coming from one of a kind guidelines. Understanding these patterns is critical for selecting the right microphone for a particular utility and for attaining the quality sound first rate in stay sound manufacturing.

1. Omnidirectional

Description:

Omnidirectional microphones pick up sound similarly from all instructions (360 ranges).

Characteristics:

No Directional Bias: Captures ambient sound and room acoustics efficaciously.

Consistent Frequency Response: Generally keeps a consistent frequency reaction from all guidelines.

Applications:

Ambiance: Recording room sound or environment.

Ensembles: Capturing sound from a set of performers.

Interviews: Multi character interviews in which the route of the sound supply adjustments.

Advantages:

Natural sound capture Easy to place

Disadvantages:

Sensitive to feedback in live settings

Less isolation from unwanted sounds

2. Cardioid

Description:

Cardioid microphones have a heart fashioned pickup pattern, with most sensitivity on the front and reduced sensitivity at the perimeters and rear.

Characteristics:

Front Focused: Captures sound normally from the front even as rejecting sound from the perimeters and rear.

Good Isolation: Provides higher isolation from heritage noise and remarks.

Applications:

Vocals: Popular for live vocal performances (e.G., Shure SM58).

Instruments: Suitable for miking character devices in a loud environment.

Live Sound: Reduces remarks by means of rejecting sound from video display units and different level assets.

Advantages:

Reduces history noise

Minimizes feedback

Good for stay and studio use

Disadvantages:

Less powerful for taking pictures ambient sound

Proximity impact (extended bass reaction whilst close to the source)

3. Supercardioid and Hypercardioid

Description:

These patterns are similar to cardioid but with a narrower pickup attitude on the front and a small lobe of sensitivity on the rear.

Supercardioid: Slightly wider than hypercardioid but nonetheless centered.

Hypercardioid: Narrowest pickup angle with extra rear sensitivity.

MIXING CONSOLES

A blending console, additionally called an audio mixer, soundboard, or mixing table, is a important piece of equipment in stay sound manufacturing. It permits the sound engineer to blend and alter multiple audio alerts, ensuring that each sound supply is balanced and clear. Understanding the various kinds and features of mixing consoles is important for effective stay sound management.

1. Types of Mixing Consoles

Analog Mixing Consoles

Description:

Analog mixers use bodily circuitry and additives to technique audio indicators.

Characteristics:

Warm Sound: Analog mixers are frequently praised for their warm and natural sound.

Hands On Control: Physical knobs and faders provide tactile remarks, making them intuitive to use.

Applications:

Live Sound: Popular in live sound settings due to their honest operation.

Studios: Used in recording studios for his or her sound exceptional and ease of use.

Advantages:
- Natural, warm sound
- Reliable and robust
- Immediate control over parameters

Disadvantages:
- Limited processing skills
- Larger and heavier than digital mixers
- Digital Mixing Consoles

Description:

Digital mixers use virtual signal processing (DSP) to deal with audio alerts.

Characteristics:

Advanced Features: Offer good sized processing alternatives, together with EQ, compression, consequences, and automation.

Recall and Presets: Settings can be saved and recalled, making an allowance for quick setup and consistency.

Applications:

Live Sound: Widely used for their flexibility and superior features.

Broadcast: Used in broadcast environments for their unique manage and integration abilities.

Advantages:

Extensive processing electricity

Compact and lightweight

Re callable settings and presets

Disadvantages:

Complexity: Can be extra complex to operate, in particular for beginners.

Learning Curve: Requires time to research and grasp digital interfaces.

POWERED MIXING CONSOLES

Description:

Powered mixers include builtin amplifiers, letting them pressure audio system directly.

Characteristics:

Convenience: Combines mixer and amplifier in a single unit, simplifying setup.

Portability: Often greater transportable and perfect for small to medium sized venues.

Applications:

Small Venues: Ideal for small clubs, houses of worship, and portable PA systems.

Rehearsals: Commonly used in band rehearsals and small stay occasions.

Advantages:

Allin one answer

Simplified setup

Cost powerful for small setups

Disadvantages:

Limited electricity output in comparison to split systems

Less flexibility in massive venues

2. Key Features of Mixing Consoles

Channels:

Mono and Stereo Channels: Each enter channel can be mono (single input) or stereo (dual enter).

Input Types: Include microphone (XLR) inputs, line degree (TRS/TS) inputs, and now and again digital inputs.

EQ (Equalization):

Basic EQ: Simple bass, mid, and treble controls.

Advanced EQ: Parametric EQs with adjustable frequency, gain, and bandwidth controls.

Aux Sends:

Monitor Mixes: Used to create separate monitor mixes for performers.

Effects Sends: Send signal to external effects processors or inner results.

Faders:

Volume Control: Linear sliders that manage the volume of every channel.

Master Fader: Controls the general output level of the mix.

Buses:

Subgroups: Group more than one channels collectively for collective processing and volume manage.

Main Bus: The very last blend output that goes to the speakers or recording device.

Effects:

Builtin Effects: Reverb, delay, chorus, and other effects are regularly protected in virtual mixers.

External Effects Integration: Analog mixers generally use outside results processors connected through aux sends.

Dynamics Processing:

Compression: Controls the dynamic range of the audio sign, making loud sounds quieter and quiet sounds louder.

Gate: Cuts off sound under a sure threshold, reducing history noise.

Routing:

Flexible Routing: Ability to path audio alerts to various outputs and buses.

Matrix Mixers: Advanced routing alternatives for complex audio setups.

3. Practical Tips for Using Mixing Consoles

Gain Staging:

Set Proper Gain Levels: Ensure each channel's input gain is ready efficiently to avoid distortion and noise.

Use PFL (PreFade Listen): Check and modify advantage tiers the use of the PFL function to screen signals before they attain the primary mix.

EQ Adjustments:

Cut Before Boosting: Start by way of slicing undesirable frequencies instead of boosting favored ones to maintain headroom and readability.

Use EQ Sparingly: Apply EQ to beautify clarity and stability with out over processing.

ANALOG VS. DIGITAL MIXERS

Choosing among analog and digital mixing consoles depends on different factors including the dimensions of the venue, the complexity of the audio setup, price range, and private options. Both types have their precise features, advantages, and barriers.

ANALOG MIXERS

Description:

Analog mixers use bodily circuitry and components to technique audio signals.

They rely upon analog sign paths and manual control interfaces.

Characteristics:

Sound Quality: Analog mixers are regarded for his or her warm, herbal sound due to the analog sign direction.

Tactile Controls: Physical knobs, faders, and switches provide direct, hand son manage over audio parameters.

Simplicity: Generally straightforward to function, with a focal point on fundamental audio controls.

Advantages:

Warm Sound: Many users recognize the rich, analog sound first rate.

Reliability: Fewer digital additives imply fewer factors of failure, often ensuing in high sturdiness.

Intuitive Operation: Easy to recognize and use, mainly for those acquainted with conventional audio system.

Disadvantages:

Limited Processing: Less processing energy in comparison to digital mixers, with fewer builtin effects and EQ alternatives.

Size and Weight: Typically large and heavier, which may be cumbersome for delivery and setup.

No Recall: Settings can not be stored and recalled, requiring guide adjustment for every setup.

Ideal For:

Live Sound: Small to medium sized venues in which simplicity and reliability are key.

Studio Work: Environments where the warm sound of analog equipment is desired.

DIGITAL MIXERS

Description:

Digital mixers use digital sign processing (DSP) to address and manner audio signals. They offer a variety of advanced capabilities and capabilities via a digital interface.

Characteristics:

Advanced Processing: Digital mixers offer large processing options, including equalization, compression, consequences, and automation.

Recall and Presets: Settings can be saved, recalled, and changed with presets, allowing for short and constant setups.

Compact Design: Typically smaller and lighter than analog mixers because of fewer physical components.

Advantages:

Extensive Features: Includes advanced functions inclusive of multi band EQ, digital consequences, and automation.

Compact and Lightweight: Easier to transport and installation, in particular for huge or visiting activities.

Flexibility: Offers flexible routing options and integration with different digital systems.

Disadvantages:

Complexity: Can be more complex to function, with a steeper gaining knowledge of curve, specially for those used to analog structures.

Digital Artifacts: Some users locate that digital mixers can introduce artifacts or lack the warmth of analog sound.

Dependence on Power: Requires a solid energy source and can be at risk of electricity screw ups or virtual glitches.

Ideal For:

Large Venues: Large scale occasions or installations wherein advanced functions and versatility are required.

Broadcast and Studio Work: Settings that benefit from virtual integration and unique manipulate over audio parameters.

BASIC CONTROLS AND FUNCTIONS OF MIXING CONSOLES

Understanding the basic controls and functions of a mixing console is critical for efficaciously dealing with audio in live sound production.

1. Input Channels

Channel Strip:

Input Gain: Adjusts the extent of the audio sign getting into the channel. Ensures the signal is robust enough without clipping.

EQ (Equalization):

High, Mid, Low: Adjusts the frequency tiers (bass, mid range, treble) to shape the sound of the input.

Parametric EQ: Offers adjustable frequency, advantage, and bandwidth for extra unique control (found on a few superior mixers).

Aux Sends: Routes a part of the sign to external outcomes processors or reveal mixes.

Pan: Controls the placement of the audio signal within the stereo subject (left/proper).

Fader: Adjusts the extent level of the audio sign for the channel. Determines how plenty of the sign is covered in the most important blend.

Mute/Solo Buttons:

Mute: Silences the channel output.

Solo: Isolates the channel for tracking, permitting the engineer to listen it without other channels.

2. Master Section

Master Fader:

Main Volume Control: Adjusts the overall quantity level of the principle mix that is despatched to the speakers or recording tool.

Bus Faders:

Subgroup or Bus Faders: Control the volume of alerts routed to unique subgroups or buses, bearing in mind collective processing and adjustment.

Auxiliary Master:

Auxiliary Sends Master: Controls the overall degree of alerts dispatched to external outcomes processors or monitor mixes.

Master EQ:

Overall Equalization: Allows for changes to the general mix's tonal balance. Found on some mixers, mainly in stay sound programs.

Meters:

Level Meters: Show the sign stage for the main blend, supporting to make sure proper benefit staging and keep away from clipping.

3. Effects and Processing

BuiltIn Effects:

Effects Send/Return: Routes alerts to and from inner or outside consequences processors. Common outcomes encompass reverb, postpone, and refrain.

Effect Controls: Adjust parameters along with mix stage, time, and depth for builtin outcomes.

Dynamics Processing:

Compressor: Controls the dynamic variety of the audio sign, making loud sounds quieter and quiet sounds louder.

Gate: Reduces or eliminates historical past noise by cutting off sounds beneath a sure threshold.

4. Routing and Connectivity

Input/Output Connectors:

Microphone Inputs (XLR): Connect microphones to the mixer.

Line Inputs (TRS/TS): Connect line degree resources along with contraptions or playback devices.

Direct Outputs:

Direct Out: Sends a sign without delay from a channel to an outside device, along with a recording device or additional processing gadget.

Groups/Buses:

Group Outputs: Routes a set of channels to a not unusual output for collective processing.

Auxiliary Sends: Routes indicators to extraordinary outputs, consisting of monitor audio system or external effects processors.

Main Outputs:

Master Outputs: Send the final combined sign to the main PA system or recording tool.

5. Monitoring

Headphone Output:

Headphone Volume Control: Adjusts the quantity level of the audio signal dispatched to headphones for tracking.

Control Room Output:

Control Room Volume: Adjusts the quantity of the signal despatched to studio monitors or control room audio system.

Talkback Mic:

Talkback Feature: Allows communication from the sound engineer to performers or different crew individuals thru a devoted microphone.

SPEAKERS AND MONITORS IN LIVE SOUND PRODUCTION

Speakers and video display units are essential additives in stay sound structures, every serving special purposes to make sure the fine audio enjoy for both performers and the target market.

1. Speakers

Description:

Speakers are used to venture sound to the target audience during stay performances. They are designed to supply high sound stress stages and cover a wide region.

Types:

Main PA Speakers:

Full Range Speakers: Designed to handle a extensive frequency range, from low bass to excessive treble. They are used to supply a balanced blend of all audio sources to the audience.

Subwoofer: Specialized speakers that manage low frequency sounds (bass). They enhance the low end of the audio spectrum and are frequently used along with full variety speakers.

Top Boxes (Main Speakers): Mounted on stands or flown from rigs, they take care of mid and excessive frequencies and offer the number one sound for the audience.

Characteristics:

Power Rating: Measured in watts, indicating the most strength the speaker can handle.

Sensitivity: Measures how efficaciously the speaker converts energy into sound. Higher sensitivity manner more output for a given input strength.

Coverage Pattern: The region over which the speaker distributes sound. It may be extensive (omnidirectional) or slender (directional), affecting how sound is projected across the venue.

Applications:

Concerts and Events: Full range audio system for principal sound reinforcement.

Outdoor Events: Weather resistant fashions for out of doors use.

Large Venues: High electricity speakers and subwoofer for large insurance.

2. Monitors

Description:

Monitors are speakers utilized by performers on level to pay attention themselves and different sound resources surely.

Types:

Stage Monitors:

Wedge Monitors: Shaped like wedges, those are located on the level ground facing the performers. They offer sound from the front of the speaker, allowing performers to listen the mix.

I Near Monitors (IEMs): Wireless systems that offer a customized audio mix immediately into the performers' ears.

Characteristics:

Monitor Mixes: Each performer typically gets a separate mix tailored to their wishes. This may also consist of their very own vocals, instruments, and different performers.

Feedback Resistance: Monitors should be placed to limit remarks and make certain clean sound.

Applications:

On Stage Monitoring: Provides performers with the necessary audio cues to perform correctly.

Personal Monitoring: IEMs are used to prevent stage noise interference and provide greater specific manipulate over the audio mix.

Advantages:

Wedge Monitors: Simple setup and effective for supplying on degree sound.

I Near Monitors: Reduces stage noise and gives nonpublic manage over audio levels.

Disadvantages:

Wedge Monitors: Can reason feedback if now not properly positioned and can be loud on degree.

I Near Monitors: Requires a wireless device and right fitting, and may be steeply priced.

Summary

Speakers:

Purpose: Project sound to the audience.

Types: Fullvariety audio system, subwoofer, and top packing containers.

Characteristics: Power rating, sensitivity, and coverage sample.

Monitors:

Purpose: Provide on degree sound for performers.

Types: Wedge monitors and inear video display units.

Characteristics: Monitor mixes, comments resistance.

Types of speakers (PA, monitors, subwoofer)

Types of Speakers in Live Sound Production

In live sound manufacturing, one of a kind varieties of speakers serve diverse features to ensure premiere sound best for both the audience and performers.

1. PA (Public Address) Speakers

Description:

PA speakers, regularly referred to as complete range or foremost audio system, are designed to assignment sound to the whole target audience. They provide a balanced representation of the audio blend,

Types:

Full Range Speakers: Capable of reproducing a huge variety of frequencies, from low bass to excessive treble. These are the number one audio system used for trendy sound reinforcement in a venue.

Powered PA Speakers: Include builtin amplifiers, removing the want for separate outside amplifiers. They are handy and might simplify setup.

Characteristics:

Frequency Range: Covers the entire spectrum of audible frequencies.

Power Handling: Rated in watts, indicating the most strength the speaker can take care of without distortion.

Coverage Pattern: Determines how sound is dispersed throughout the venue. Common patterns include slim (for focused coverage) or huge (for broader insurance).

Applications:

Concerts: Provide the main sound reinforcement for song and vocals.

Events and Conferences: Ensure clean and even sound distribution for speeches and shows.

2. Monitors

Description:

Monitors are audio system used on stage to offer performers with a clean and personalized blend of the audio. They allow performers to listen themselves and different audio sources at some point of a overall performance.

Types:

Wedge Monitors: Shaped like wedges and positioned at the stage floor, going through the performers. They deliver sound immediately to the performers and are usually used in stay performances.

Floor Monitors: Similar to wedge monitors however may have exclusive shapes and designs.

I Near Monitors (IEMs): Wireless structures that provide a customized audio blend immediately into the performers' ears. They provide higher isolation from level noise and may be adjusted for a clearer mix.

Characteristics:

Monitor Mixes: Allows performers to receive a tailor made blend of audio, together with their very own vocals and gadgets.

Feedback Resistance: Proper positioning is vital to keep away from comments and make certain clear sound.

Applications:

Live Performances: Ensures performers can pay attention themselves and other critical audio elements.

Rehearsals: Provides performers with a regular and controlled audio mix.

3. Subwoofer

Description:

Subwoofer are specialized speakers designed to breed low frequency sounds (bass). They take care of frequencies under the ones produced by means of full variety speakers and decorate the general intensity and effect of the audio.

Types:

Passive Subwoofer: Require an outside amplifier to drive them. They do now not have integrated amplification.

Powered Subwoofer: Include builtin amplifiers, simplifying setup through integrating strength and speaker into one unit.

Characteristics:
- Frequency Range: Focuses on very low frequencies, generally below one hundred Hz.
- Power Handling: Rated in watts, indicating the most power the subwoofer can manage with out distortion.
- Size and Placement: Larger and heavier in comparison to complete range speakers. Proper placement is vital for most beneficial bass reaction and integration with the main speakers.

Applications:
- Concerts and Events: Enhances the low give up frequency reaction, including intensity and impact to the audio.
- Theaters and Cinemas: Provides the rumble and impact wished for dramatic sound results and movie rankings.

Monitors:

Purpose: Provide onstage sound for performers.

Types: Wedge video display units, floor video display units, inear video display units.

Characteristics: Personalized display mixes, feedback resistance.

Subwoofer:

Purpose: Reproduce low frequency sounds (bass).

Types: Passive and powered subwoofer.

Characteristics: Low frequency response, strength handling, size and placement concerns.

PLACEMENT AND SETUP OF SPEAKERS IN LIVE SOUND PRODUCTION

Proper placement and setup of speakers are critical for attaining most fulfilling sound first rate and coverage.

1. Placement of PA Speakers

Main PA Speakers:

Front of House (FOH) Placement:

Height: Position the speakers at a height that allows sound to cover the target market efficaciously. This commonly approach mounting them on stands or rigging them above the audience to avoid obstructions.

Angle: Tilt the audio system downward closer to the target market. This guarantees that sound reaches the listeners at various distances and heights and allows reduce reflections from the ceiling or top walls.

Spacing: Place the audio system equidistant from the middle of the stage to gain balanced sound distribution. The spacing will depend upon the venue length and speaker coverage patterns.

Alignment: Ensure that the audio system are aligned to cowl the whole audience place. Use a mixture of principal speakers and possibly delay speakers for larger venues to make sure constant sound.

Coverage Pattern:

Horizontal Coverage: Check that the horizontal dispersion of the speakers covers the whole target audience place. Wide dispersion audio system can cowl greater vicinity, but in large venues, more than one speakers is probably wanted.

Vertical Coverage: Adjust the vertical angle to make certain that sound reaches both the the front and rear sections of the audience. This would possibly involve angling the speakers downward or using extraordinary speaker arrays.

Considerations:

Avoiding Feedback: Position the audio system and microphones to decrease remarks. This often approach placing monitors in the front of the audio system and using directional microphones.

Power and Cables: Ensure that energy and sign cables are nicely controlled and secured to save you tripping hazards and interference.

2. Placement of Monitors

Stage Monitors:

Wedge Monitors:

Position: Place wedge monitors at the level floor dealing with the performers. Position them near the performers' locations to make sure they can hear themselves and other audio resources truly.

Angle: Tilt the video display units upward towards the performers' ears to ensure right sound projection and decrease feedback.

Avoiding Feedback: Monitor placement ought to be executed carefully to keep away from remarks with the primary PA speakers. Position them in which they're least likely to pick up sound from the primary speakers.

I Near Monitors (IEMs):

Transmitter Placement: Place the IEM transmitter in a place in which it can broadcast a clear sign to all performers. Ensure it's miles properly tuned and avoids interference.

Personal Mixes: Set up customized mixes for every performer to ensure they pay attention what they need. IEMs offer isolation from level noise and allow for more control over audio levels.

3. Placement of Subwoofer

Subwoofer:

Positioning:

Front of House (FOH): Place subwoofer near the principle PA audio system, regularly on the front of the stage or on the perimeters. This placement allows in integrating the low frequencies with the relaxation of the audio.

Ground Coupling: Position subwoofer at the floor to maximize bass reaction. They can be positioned directly at the floor or accelerated barely on stands, depending at the venue and subwoofer design.

Spacing: If using a couple of sub woofers, area them correctly to avoid segment cancellation and make sure even bass coverage. Experiment with placement to discover the high quality position for even bass reaction all through the venue.

Considerations:

Phase Alignment: Ensure subwoofer are phase aligned with the main PA speakers to avoid cancellation and make certain clean integration of low frequencies.

Avoiding Boominess: Adjust the subwoofer placement to avoid immoderate bass buildup in sure regions of the venue. This may additionally contain trial and error to locate the best position.

Monitors:

Wedge Monitors: Place on degree, angled towards performers.

I Near Monitors: Set up transmitters and personal mixes for performers.

Subwoofer:

Positioning: Place close to most important audio system, ground coupled or multiplied as wished.

Spacing and Alignment: Ensure proper phase alignment and even bass coverage.

CABLES AND CONNECTORS IN LIVE SOUND PRODUCTION

Cables and connectors are crucial components of any stay sound gadget, serving because the lifeline for transmitting audio alerts between various devices. Understanding the kinds of cables and connectors, their makes use of, and quality practices for dealing with them is crucial for reliable and notable sound manufacturing.

1. Types of Cables

1.1. XLR Cables

Description:

XLR cables are the standard for balanced audio connections, commonly used for microphones, expert audio device, and some line level connections.

Features:

Balanced Signal: XLR cables bring a balanced signal, which enables lessen noise and interference, making them perfect for lengthy cable runs.

Three Pin Configuration: Typically have three pins (superb, bad, and ground), making sure a stable and noise loose connection.

Uses:

Microphones: Connecting microphones to mixers or audio interfaces.

DI Boxes: Connecting gadgets to mixers thru direct input (DI) bins.

Speakers: Used in powered speakers and a few PA structures.

1.2. TRS (Tip Ring Sleeve) Cables

Description:

TRS cables are used for balanced audio connections, similar to XLR, and can also be used for stereo indicators.

Features:

Balanced or Stereo: Depending on the application, TRS cables can bring either a balanced mono sign or an unbalanced stereo sign.

1/4Inch or 1/8Inch: Common sizes include 1/four inch (used for professional audio) and 1/eight inch (used for purchaser audio devices).

Uses:

Instruments: Connecting instruments like keyboards and electric guitars to mixers or amplifiers.

Headphones: Connecting headphones to audio interfaces or mixers.

Insert Points: Used for connecting outside processors like compressors or EQs to mixers.

1.3. TS (TipSleeve) Cables

Description:

TS cables are unbalanced and typically used for mono connections.

Features:

Unbalanced Signal: Carries an unbalanced sign, which is greater liable to noise and interference, so they may be exceptional for quick cable runs.

1/four Inch Size: Commonly utilized in units and line level system.

Uses:

Instruments: Connecting electric guitars or basses to amplifiers or pedal boards.

Line Level Equipment: Connecting unbalanced line degree device to mixers or amplifiers.

1.4. Speaker Cables

Description:

Speaker cables are designed to carry high electricity audio indicators from amplifiers to passive audio system.

Features:

Thicker Gauge: Typically have a thicker gauge (decrease AWG number) to handle better present day with out signal loss.

Unshielded: Unlike tool or mic cables, speaker cables are unshielded, as the better sign ranges are much less susceptible to interference.

Uses:

Passive Speakers: Connecting passive audio system to amplifiers.

Subwoofer: Connecting subwoofer to amplifiers.

1.5. RCA Cables

Description:

RCA cables are commonly used for unbalanced connections in customer audio and some professional applications.

Features:

Color Coded: Typically purple and white (or black) connectors for stereo audio, with yellow frequently used for video.

Unbalanced Signal: Carries unbalanced audio, making them greater appropriate for quick runs in less crucial packages.

CHAPTER 3: SETTING UP A LIVE SOUND SYSTEM

STAGE LAYOUT

POSITIONING MICROPHONES AND SPEAKERS IN LIVE SOUND PRODUCTION

Proper positioning of microphones and audio system is vital for reaching clean, balanced sound and minimizing issues like comments, segment cancellation, and unwanted noise.

1. Positioning Microphones

1.1. Vocal Microphones

On Stage Placement:

Distance from Performer: Place the microphone approximately 612 inches faraway from the performer's mouth for clear vocal capture. Closer placement will increase the proximity impact (improving bass), even as in addition distance reduces this effect.

Angle: Position the microphone at a slight upward or downward angle to minimize plosive sounds (like "p" and "b" sounds) and reduce the chance of comments.

Stability: Use a sturdy microphone stand to maintain the microphone in place, specifically all through dynamic performances.

Avoiding Feedback:

Monitor Placement: Position wedge video display units in front of the microphone, angled toward the performer, to decrease comments. The microphone have to be pointed far from the screen.

Directional Microphones: Use cardioid or supercardioid microphones to consciousness at the performer's voice and reject sound from the sides and rear, lowering remarks threat.

1.2. Instrument Microphones

Guitar Amplifiers:

 Close Miking: Place the microphone 1three inches from the amplifier's speaker cone, barely off middle, for a balanced tone. Closer placement enhances low frequencies, even as off middle placement captures a brighter sound.

 Angle: Angle the microphone slightly toward or far from the speaker cone to alter the tonal stability.

Drums:

 Snare Drum: Place the microphone 1three inches above the drum head, angled closer to the center of the drum, to seize the assault and body of the snare.

 Bass Drum: Position a microphone within the drum, close to the beater for a punchy sound, or similarly back for a extra resonant tone.

Overheads: Place overhead microphones 2three ft above the drum kit, spaced equally on both facet, to capture the cymbals and universal kit sound.

Acoustic Instruments:

Guitars and Strings: Place the microphone 612 inches from the sound hole or frame, angled barely to seize a herbal, balanced tone.

Pianos: Use a pair of microphones, one near the low strings and one near the excessive strings, for a complete variety sound. Adjust the height and attitude to balance the bass and treble.

1.3. Ambient Microphones

Room Miking:

Distance from Source: Position ambient microphones similarly away from the sound supply (1020 feet) to capture the natural reverb and room acoustics.

Height and Angle: Place microphones at ear stage or better, angled slightly downward to pick up the general sound environment.

AVOIDING FEEDBACK IN LIVE SOUND PRODUCTION

Feedback happens while the sound from speakers reenters a microphone, gets amplified, and is then picked up by means of the identical microphone, creating a loop that effects in a high pitched squeal. Proper setup and positioning can extensively lessen the hazard of feedback.

1. Microphone Placement and Selection

1.1. Use Directional Microphones

Cardioid or Supercardioid Microphones: These microphones pick up sound commonly from the the front and reject sound from the perimeters and rear. This makes them less probable to select up sound from video display units or PA speakers.

Hypercardioid Microphones: These provide even narrower pickup patterns, in addition lowering the chances of feedback, but require extra unique positioning.

1.2. Positioning Microphones

Distance from Sound Sources: Keep microphones near the sound supply (e.G., vocalist or device) to boom the sign to noise ratio, because of this you can hold the advantage lower and decrease the chance of comments.

Avoid Direct Alignment with Speakers: Ensure that the microphone isn't directly aligned with any speakers. The again of the microphone need to face the audio system, as this is the vicinity where directional microphones are least sensitive.

Monitor Placement: Position stage video display units to the edges of performers, angled away from the microphones. This reduces the chance of sound from the video display units being picked up via the microphone.

1.3. Use of Multiple Microphones

 Minimize the Number of Open Microphones (NOM): The extra microphones which might be lively, the more the chance of remarks. Mute microphones after they're not in use, and use a minimal wide variety of microphones necessary to capture the sound.

 2. Speaker Placement

2.1. Main PA Speakers

 Distance from Microphones: Place PA audio system properly in the front of the microphones to limit the hazard of the microphones selecting up sound from the audio system.

 Height and Angle: Elevate PA speakers above the level of the microphones and target audience to project sound over their heads, reducing the chance that sound will loop lower back into the microphones.

2.2. Monitor Placement

Angle and Distance: Position degree monitors near the performers but angled so that the sound is directed toward their ears and faraway from the microphone.

Use Multiple Monitors: For large degrees, use extra monitors at lower volumes instead of fewer monitors at better volumes. This reduces the general sound stress and capacity feedback.

3. Sound System Settings

3.1. Gain Structure

Proper Gain Staging: Set the benefit degrees on your mixer successfully. Start with the aid of setting the advantage on each channel so that the enter stage is powerful but not peaking. This lets in you to hold lower fader degrees, lowering the general volume and the chance of comments.

Use Pad Switches: Engage the pad switch on the mixer or microphone if the sign is simply too warm, which reduces the input level and enables control remarks.

3.2. Equalization (EQ)

Feedback Frequencies: Identify and cut the frequencies which might be prone to feedback. These are commonly within the mid range (round 1kHz to 3kHz) or excessive frequencies. Use a parametric EQ or a image EQ to notch out these frequencies.

High Pass Filters: Apply high pass filters on vocal and tool channels to lessen low frequency rumble that could lead to feedback.

3.3. Use of Feedback Suppression Tools

Automatic Feedback Suppressors: These devices discover and decrease remarks frequencies robotically through applying slender band EQ cuts. They can be beneficial in complex setups where guide EQ modifications are difficult.

Limiter and Compressor Settings: Set compressors and limiters cautiously, as over compression can cause remarks via decreasing the dynamic range, making quieter sounds (which include ability feedback loops) greater audible.

4. Sound Check and Monitoring

4.1. Sound Check

Walk the Venue: During sound take a look at, stroll around the venue to pay attention for ability remarks spots. Make adjustments to speaker placement, microphone positioning, and EQ settings as needed.

Incremental Volume Increases: Increase quantity progressively at the same time as monitoring for comments. Identify hassle frequencies and address them earlier than the performance starts off evolved.

SIGNAL FLOW IN LIVE SOUND PRODUCTION

Signal glide refers back to the route that an audio sign takes from the supply (like a microphone or tool) through various stages of processing and amplification, till it ultimately reaches the audio system.

1. Basic Components of Signal Flow

1.1. Source (Input Devices)

Microphones: Convert acoustic sound into an electrical sign. Different types (dynamic, condenser, and so forth.) are used relying at the sound supply.

Instruments: Electric guitars, keyboards, and other gadgets output electric indicators that need amplification and processing.

DI Boxes: Convert high impedance, unbalanced alerts (like from electric guitars or keyboards) into low impedance, balanced indicators, which might be much less at risk of noise and can journey longer distances with out degradation.

1.2. Stage Box/Snake

Stage Box: A centralized hub on stage where all microphones and gadgets are linked. From here, a multicore cable (regularly called a "snake") carries the alerts to the mixing console.

1.3. Mixing Console

Preamp: The first prevent for the sign on the console, wherein it's miles amplified to a level appropriate for processing. Proper advantage staging is critical right here to avoid noise or distortion.

Channel Strip: Each enter has a channel strip at the console, which includes EQ, faders, panning, and aux sends. Here, the sign can be shaped, balanced with different alerts, and routed to numerous locations.

Aux Sends/Returns: Used to send indicators to external results processors or stage video display units. The processed signal can then be lower back to the console and blended again into the principle output.

Groups/Subgroups: Channels may be grouped together (e.G., all drum mics) and controlled with a unmarried fader. This simplifies blending and allows for greater cohesive processing.

Master Output: The blended mix of all channels is sent thru the master output fader, which controls the overall quantity of the mix being despatched to the speakers.

1.4. Outboard Gear (Optional)

Equalizers (EQs): Further refine the tone of the sign, frequently used to notch out remarks frequencies or shape the general sound.

Compressors/Limiters: Control the dynamic variety of the signal, making quiet sounds louder and loud sounds quieter. This ensures a extra consistent output degree.

Effects Processors: Add reverb, put off, chorus, or other effects to decorate the sound.

1.5. Amplifiers

Power Amplifiers: Amplify the line level sign from the integration console to a stage which can drive speakers. Some modern day audio system are powered, which means the amplification is integrated, disposing of the need for external amplifiers.

1.6. Speakers (Output Devices)

Main PA Speakers: Project the final mix to the target market.

Monitors: Provide performers with their personal mix on level, permitting them to listen themselves and different vital elements of the performance.

Subwoofer: Reproduce low frequency sounds, adding depth and electricity to the general sound.

2. Detailed Signal Flow Example

2.1. Microphone to Mixing Console

1. Microphone captures the sound and converts it to an electrical sign.
2. XLR Cable incorporates the balanced sign to the stage container.
3. Stage Box/Snake consolidates more than one signals and sends them to the combination console.
4. Mixer's Preamp amplifies the sign to line stage.
5. Channel Strip Processing (EQ, compression, and so on.) shapes the signal.

2.2. Processing and Routing

6. Aux Sends path a part of the signal to outcomes processors or level monitors.
7. Insert Points allow external processors (e.G., compressors, EQs) to modify the signal.
8. Subgroups integrate associated signals (e.G., all drum mics) right into a unmarried fader for less complicated manipulate.

2.3. Output to Amplifiers and Speakers

9. Master Output sends the final blend to the power amplifiers.
10. Power Amplifiers boost the signal to speaker level.
11. Speakers (PA) broadcast the sound to the target market.
12. Monitors offer a separate mix for performers.

3. Common Variations in Signal Flow

3.1. Digital Mixing Consoles

Digital Signal Processing (DSP): In a virtual mixer, most processing (EQ, compression, results) is achieved digitally within the console itself.

Networked Audio: Some systems use digital networks (e.G., Dante) to send audio alerts over Ethernet cables, reducing the want for traditional analog snakes.

3.2. I Near Monitor (IEM) Systems

Transmitter: The blend is sent to a wifi transmitter.

IEM Receiver: Performers wear receivers that choose up the sign and ship it to their inear video display units, offering a personalized and isolated blend.

UNDERSTANDING THE SIGNAL PATH FROM INPUT TO OUTPUT IN LIVE SOUND PRODUCTION

The signal direction refers to the adventure an audio signal takes from its starting place (enter) to its very last vacation spot (output) in a live sound setup. Understanding this path is critical for setting up,

1. Input Stage

1.1. Sound Source

Microphones: Convert acoustic sound (voice, contraptions) into electrical signals. The type of microphone (dynamic, condenser, and many others.) and its placement affect the fine and traits of the captured sound.

Instruments: Electric guitars, keyboards, and other digital gadgets produce line level signals that may be directly input into the machine or handed through a Direct Input (DI) container.

1.2. Direct Input (DI) Box

Function: Converts excessive impedance, unbalanced alerts from devices into low impedance, balanced indicators. This conversion reduces noise and allows the sign to travel longer distances without degradation.

1.3. Cables

XLR Cables: Used for microphones and balanced signals, minimizing noise and interference.

Instrument Cables: Typically 1/4" TS (Tip Sleeve) cables for unbalanced alerts from devices.

2. Pre Amplification

2.1. Preamp Stage

Mixing Console Preamps: The first forestall within the signal route, where the audio sign is amplified to a usable level (line level). Proper advantage staging is crucial to keep away from noise and distortion.

External Preamps (Optional): Sometimes used to acquire a specific sound person or for higher exceptional amplification.

3. Signal Processing

3.1. Channel Strip

- EQ (Equalization): Adjusts the tonal stability of the sign by way of boosting or slicing precise frequency tiers.
- Compression: Controls the dynamic range of the signal, making quiet sounds louder and loud sounds quieter.
- Panning: Determines the sign's position inside the stereo field (left, right, or center).

Fader: Controls the sign stage going to the combination bus. The fader is frequently used for typical quantity adjustment in the course of the live blend.

3.2. Inserts and Aux Sends

Inserts: Allow external processors (e.G., compressors, gates) to be placed directly in the sign path for character channel processing.

Aux Sends: Route a part of the signal to external effects devices (e.G., reverb, put off) or to stage video display units for performers. The processed sign can then be lower back to the mixture via aux returns or committed go back channels.

3.3. Grouping and Subgroups

Subgroups: Combine a couple of channels (e.G., all drum mics) into a unmarried group for collective processing and manipulate. This allows for simpler management and regular processing throughout related indicators.

4. Mixing and Master Output

4.1. Master Bus

Final Mix: The indicators from all channels are mixed into the master bus. The master fader controls the overall output degree despatched to the energy amplifiers or powered speakers.

Main Processing: Additional EQ, compression, and limiting may be implemented at the grasp bus to manipulate the general mix and ensure it's inside safe levels for the audio system.

4.2. Power Amplifiers (for Passive Speakers)

Function: Amplify the road degree signal from the integration console to a degree that could force passive speakers. The preference of amplifier relies upon on the energy necessities of the speakers.

5. Output Stage

five.1. Speakers

PA (Public Address) Speakers: Project the final blend to the audience. These can be passive (requiring an outside amplifier) or lively (with integrated amplifiers).

Monitors: Provide performers with their own mix on level, permitting them to pay attention themselves and different key factors. This helps performers live in sync with the music and each different.

Subwoofers: Reproduce lowfrequency sounds, adding depth and electricity to the general sound. They are generally located at the ground or rigged to reinforce the bass frequencies.

5.2. Final Output Considerations

Speaker Placement: Proper placement and alignment of speakers make certain even sound insurance and reduce segment troubles or remarks.

CHAPTER 4: SOUND CHECK AND LINE CHECK

IMPORTANCE OF SOUND CHECK

GOALS OF A SOUND CHECK IN LIVE SOUND PRODUCTION

A sound test is an vital step in live sound production, conducted earlier than a overall performance to make certain that the sound system is nicely set up and optimized. The primary dreams of a legitimate test are to make certain audio best, stability, and clarity, in addition to to cope with any capacity troubles before the overall performance starts off evolved.

1. Set Up and Test All Equipment

1.1. Verify Equipment Functionality

Check All Connections: Ensure that each one microphones, contraptions, cables, and devices are connected efficaciously and functioning as predicted.

Test Input Signals: Confirm that every microphone and instrument is generating a easy, strong sign without noise or distortion.

Monitor Speakers and Monitors: Verify that each one PA audio system, video display units, and subwoofer are working successfully and are positioned properly.

1.2. Confirm Signal Flow

Trace the Signal Path: Ensure the signal is travelling effectively from the enter (microphone/instrument) via the integration console, processing gadget, amplifiers, and out to the speakers.

Check for Latency or Delays: Make sure there is no significant postpone between the enter and output signals, which could disrupt the overall performance.

2. Set Proper Gain Structure

2.1. Optimize Gain Levels

Set Preamp Gain: Adjust the gain on each channel to make sure that alerts are robust sufficient for processing with out clipping or distortion.

Headroom: Allow enough headroom to accommodate louder elements of the overall performance with out causing distortion.

2.2. Balance Input Levels

Match Levels Across Channels: Ensure that all input channels (vocals, gadgets) have appropriate levels relative to every other to prevent one supply from overpowering others.

Avoid Feedback: Set gain degrees and regulate microphone placement to decrease the threat of feedback.

3. Adjust EQ and Processing

3.1. Equalize Channels

Tailor the Sound: Adjust the EQ on every channel to decorate the herbal sound of the vocals and contraptions, getting rid of unwanted frequencies and emphasizing the desired tonal traits.

Notch Out Problem Frequencies: Identify and decrease frequencies susceptible to comments or muddiness.

3.2. Apply Compression and Effects

Dynamic Control: Set up compressors to manage the dynamic range of vocals and contraptions, ensuring consistent ranges in the course of the overall performance.

Effect Levels: Adjust reverb, put off, and different results to combo evidently with the overall performance without overwhelming the direct sound.

4. Balance the Mix

4.1. Create a Cohesive Mix

Balance Front of House (FOH) Mix: Ensure that each one elements of the performance (vocals, instruments, backing tracks) are balanced and blend properly together within the essential mix.

STEPS TO PERFORM A SOUND CHECK IN LIVE SOUND PRODUCTION

A properly achieved sound take a look at is essential for making sure a easy performance. It involves systematic testing and adjustment of the sound system to reap most reliable sound first class and stability.

1. Preparation

1.1. Set Up Equipment

Position and Connect All Equipment: Ensure that each one microphones, instruments, monitors, PA speakers, and different device are successfully placed and linked.

Label Channels: Label every channel on the mixing console for clean identity for the duration of the sound check (e.G., "Lead Vocal," "Guitar 1," "Kick Drum").

Power On: Turn on all device, which includes the integration console, amplifiers, and outboard tools. Ensure the whole lot is in desirable operating order.

1.2. Communicate with Performers

Coordinate with the Band: Ensure all performers are geared up and privy to the sound check process. Clear communique is key to an green sound check.

Explain the Process: Briefly provide an explanation for the sound test steps to the performers, so that they recognize what to anticipate and what you want from them.

2. Check and Set Gain Levels

2.1. Start with a Basic Sound Source

Test Individual Channels: Have each performer play or sing one by one. Start with the kick drum or bass guitar, as they typically offer the inspiration for the mixture.

Set Preamp Gain: Adjust the benefit on each channel so that the signal is strong but not clipping. Aim for a level that peaks just underneath 0dB on the meter.

2.2. Work Through All Inputs

Proceed to Each Instrument/Vocal: Work methodically through all inputs, putting advantage stages and ensuring each source is clear and at the proper volume.

Check for Feedback: As you place advantage levels, concentrate for comments and adjust microphone placement or advantage settings to remove it.

3. Adjust Equalization (EQ)

3.1. EQ Individual Channels

Enhance Tonal Balance: Adjust the EQ for every channel to beautify the herbal sound of the tool or voice. Remove any undesirable frequencies that motive muddiness or harshness.

Address Room Acoustics: Make EQ modifications to account for the acoustics of the venue, which could affect how certain frequencies are perceived.

3.2. Focus on Problem Frequencies

Notch Out Feedback: If feedback happens, discover the intricate frequency and reduce it the use of the EQ.

Fine Tune the Mix: Continue refining the EQ for each channel till the general mix is balanced and clear.

4. Check and Adjust Monitor Mixes

4.1. Set Monitor Levels

Ask for Performer Feedback: Have every performer play or sing at the same time as you adjust their monitor mix. Ensure they can pay attention themselves and some other important devices or vocals.

Balance Monitor Mixes: Create a balanced mix for each monitor or inear monitor machine, catering to the desires of person performers.

4.2. Ensure Clarity Without Feedback

Monitor Placement: Ensure video display units are positioned efficiently to avoid feedback, with cautious attention to the perspective and distance from microphones.

Adjust Monitor EQ: If wished, alter the EQ at the monitors to lessen the threat of remarks whilst preserving clarity for the performers.

5. Balance the Front of House (FOH) Mix

5.1. Bring in the Full Band

Play a Full Song: Have the complete band carry out a music to check the overall balance. This offers you an opportunity to pay attention how all of the factors have interaction within the mix.

Adjust Faders: Make modifications to the fader tiers for every channel to acquire a balanced and cohesive blend that sounds good in the audience location.

5.2. Refine the Mix

Focus on Clarity: Ensure that vocals are clear and distinguished, and that no device or vocal is overpowering others.

Consider Dynamics: Adjust compression settings if essential to manipulate dynamics and ensure a steady sound in the course of the overall performance.

6. Test Effects and Final Adjustments

6.1. Apply Effects

Check Reverb, Delay, and Other Effects: Test any results you propose to apply throughout the performance. Adjust the ranges to ensure they supplement the mixture with out overwhelming it.

CHAPTER 5: BALANCING THE MIX

UNDERSTANDING THE MIX

COMPONENTS OF A STAY MIX (VOCALS, CONTRAPTIONS, CONSEQUENCES)

A live blend is the stability of numerous audio factors at some point of a live overall performance, inclusive of vocals, devices, and outcomes. Each component needs to be carefully managed to make sure readability, stability, and effect for the target audience.

1. Vocals

1.1. Lead Vocals

Prominence in the Mix: Lead vocals are commonly the maximum important detail in a stay mix, as they convey the main melody and lyrics. They ought to be clear, intelligible, and sit down prominently above the devices.

EQ: Adjust EQ to enhance clarity and decrease muddiness, usually by slicing low mid frequencies and boosting presence within the mid excessive range (round 2kHz to 5kHz).

Compression: Apply compression to control dynamic range, making sure the vocals stay constant and do not wander off in the mix all through quieter passages.

1.2. Backing Vocals

Blend with Lead Vocals: Backing vocals ought to complement and help the lead vocals without overpowering them. They are usually positioned barely decrease within the mix.

Panning: Pan backing vocals slightly to the left or proper to create a experience of area and separation from the lead vocals.

EQ and Effects: Apply comparable EQ and compression as the lead vocals, with diffused reverb or delay to combination them into the general blend.

2. Instruments

2.1. Rhythm Section

Kick Drum: Provides the low quit basis of the mix. It have to be punchy and clear, regularly boosted around 60Hz to 100Hz for effect and around 3kHz to 5kHz for assault.

Bass Guitar: Works with the kick drum to create a solid lowend. It should be clean and present without overwhelming the mix. Boost within the low mid range (round 80Hz to 200Hz) to make certain it's felt as well as heard.

Snare Drum: Adds rhythm and snap to the mixture. It generally sits round 200Hz to 250Hz for frame and 5kHz to 10kHz for snap and brightness.

2.2. Harmonic Instruments

Guitars: Electric and acoustic guitars add harmonic richness and texture. Electric guitars are often panned left and right for stereo width. EQ modifications would possibly encompass slicing low mids to avoid muddiness and boosting round 3kHz to 5kHz for presence.

Keyboards/Pianos: Provide melodic and harmonic assist. They frequently sit down within the mid range of the combination. Depending on their function, they is probably panned barely left or proper to create space.

2.3. Percussion

HiHats and Cymbals: Provide high frequency sparkle and rhythmic detail. They must be crisp and clean, regularly boosted around 5kHz to 10kHz. These are commonly panned barely to feature width to the mixture.

Additional Percussion: Instruments like congas, bongos, or shakers upload rhythmic complexity. Their placement within the mix relies upon on their function however regularly they're panned to create a greater immersive enjoy.

3. Effects

3.1. Reverb

Adds Depth and Space: Reverb creates a sense of space and might make the combination sound extra herbal. It may be applied to vocals, contraptions, or the complete mix relying on the desired effect.

Types of Reverb: Plate reverb is often used on vocals for a easy, polished sound, even as hall or room reverb might be used on drums or guitars to create a sense of area.

3.2. Delay

Creates Echo and Rhythmic Interest: Delay outcomes can add rhythmic repeats to vocals or units, growing a experience of intensity and motion. Short delays can thicken a sound, while longer delays can create echo results.

Application: Delay is regularly used sparingly on lead vocals to enhance certain phrases or to create a more spacious blend.

3.Three. Modulation Effects

Chorus, Flanger, Phaser: These consequences are used to add richness and movement to instruments, in particular guitars and keyboards. They could make a legitimate wider and more complex.

Application: Modulation results are commonly used subtly to avoid cluttering the mix, supplying a lush or dreamy great to the sound.

3.4. Compression and Dynamics Processing

Control Dynamics: Compression is used on most elements of a live mix to govern dynamic variety, making sure that the levels stay constant. This is particularly essential in a stay placing wherein performance dynamics can vary greatly.

Multi band Compression: In a few cases, multi band compression is used to manipulate specific frequency stages, making an allowance for greater unique dynamic control.

3.5. Gates and Expanders

Noise Control: Gates are used to put off undesirable noise or bleed from different gadgets, particularly on drums. They open whilst the signal exceeds a sure threshold and near while it falls underneath it.

Tightening the Mix: Gates can assist tighten the sound of drums by using casting off bleed from different drum mics, creating a purifier, greater focused sound.

4. Overall Mix Balance

4.1. Volume Levels

Achieve Balance: Adjust fader levels to make certain that all components (vocals, contraptions, results) are balanced with regards to each different. No single element need to overpower the combination except intentionally featured.

Dynamic Mixing: During the overall performance, the mixture may additionally want to be adjusted in real time to respond to adjustments in the energy or dynamics of the overall performance.

4.2. Panning

Stereo Image: Use panning to create a extensive stereo photograph, giving the combination depth and area. Vocals are normally targeted, at the same time as instruments may be panned to distinctive positions inside the stereo subject.

Avoid Overcrowding: Careful panning enables keep away from overcrowding in the center of the mixture and guarantees clarity and separation of factors.

4.3. Master Bus Processing

Final EQ and Compression: Apply final EQ and compression to the grasp bus to glue the mix collectively, making sure it sounds cohesive and polished.

Limiter: A limiter may be used at the grasp bus to prevent clipping and protect the speakers from sudden extent spikes.

BALANCING LEVELS IN A LIVE MIX

Balancing stages is one of the most important aspects of stay sound manufacturing. It ensures that each one factors of the performance—vocals, devices, and effects—are heard without a doubt and harmoniously, without any unmarried element overpowering the others. Here's the way to efficaciously stability ranges in a live mix:

1. Initial Setup

1.1. Set a Reference Level

Start with a Foundation: Begin by using putting a reference level for one of the foundational elements of the combination, typically the kick drum or bass guitar. This helps establish the low quit of the combination and serves as a foundation for balancing other elements.

Use the Mixing Console's Meters: Keep a watch on the meters to ensure the degrees are sturdy but now not peaking. Aim for peaks just underneath 0dB, allowing for headroom.

1.2. Bring in Each Element Sequentially

Add Vocals First: After putting the muse, bring within the lead vocals. Adjust the fader so the vocals are truly heard over the kick drum or bass guitar.

Introduce Instruments Gradually: One by way of one, bring in each device, adjusting their tiers in order that they complement rather than compete with the vocals and foundational elements.

2. Achieving Balance

2.1. Balance Rhythm Section

Kick Drum and Bass: Ensure those factors are balanced with every different. The bass should fill out the low end with out muddying the kick drum. Adjust EQ if vital to create area for each.

Snare Drum and HiHats: Balance the snare so it cuts thru the mix, supplying rhythm and snap, while hihats ought to add sparkle with out overpowering different factors.

2.2. Balance Harmonic Instruments

Guitars and Keyboards: Balance guitars and keyboards towards each other and the rhythm phase. Use panning to separate them within the stereo area, which enables in growing readability and space.

Avoid Masking: Make sure no unmarried instrument masks some other. For example, avoid having the guitar frequencies overlap too much with vocals, that may result in muddiness.

2.3. Blend Vocals and Instruments

Lead Vocals: Keep the lead vocals barely above the opposite factors inside the blend, as they are normally the point of interest. Backing vocals should be audible however supportive, no longer competing with the lead.

Monitor Levels: Continuously monitor vocal ranges in the course of the overall performance, especially if the singer varies their dynamics. Adjust in real time as wished.

3. Fine Tuning with EQ and Compression

3.1. Use EQ for Clarity

Cutting vs. Boosting: Use EQ cuts to carve out area for each element, instead of simply boosting frequencies. For instance, slicing low mids in guitars can make room for the vocals.

Avoid Frequency Clashes: Identify and reduce frequencies that clash among units, such as the low mid variety that regularly competes among guitars, keyboards, and vocals.

3.2. Apply Compression

Control Dynamics: Use compression to control the dynamic variety of individual elements, ensuring they stay consistent within the mix without surprising extent spikes.

Group Compression: Consider the use of compression on subgroups (e.G., all drums) to achieve a greater cohesive sound with out over compressing person elements.

4. Panning for Stereo Balance

4.1. Create a Stereo Image

Panning Instruments: Pan devices like guitars, keyboards, and percussion slightly left or right to create a extensive stereo photograph. This allows save you overcrowding in the middle and gives each detail its personal space.

Center Important Elements: Keep key elements like lead vocals, bass, and kick drum targeted to keep stability and impact.

4.2. Stereo Effects

Use Effects Wisely: Apply reverb and delay outcomes in stereo to create a experience of space, however make certain they don't wash out the mixture. Adjust the moist/dry stability so the results enhance rather than overpower the original signal.

5. Final Adjustments

5.1. Walk the Venue

- Check from Different Locations: Walk around the venue to listen to the mixture from diverse positions. Ensure that the

tiers are balanced throughout the target market place and modify as wanted.
- Adjust for Venue Acoustics: Take into consideration the venue's acoustics, which could have an effect on how sound ranges are perceived in one of a kind components of the space. Make modifications to maintain clarity and stability.

www.ingramcontent.com/pod-product-compliance
Lightning Source LLC
Chambersburg PA
CBHW071932210526
45479CB00002B/654